INTERNATIONAL CENTRE FOR MECHANICAL SCIENCES

COURSES AND LECTURES - No. 253

GIOVANNI MANFRE'
TECHNION SPA
NOVARA (ITALY)

LIMIT OF THE SPINNING PROCESS IN MANUFACTURING SYNTHETIC FIBERS

COURSE HELD AT THE DEPARTMENT
OF GENERAL MECHANICS

UDINE 1975

SPRINGER-VERLAG WIEN GMBH

This work is subject to copyright.

All rights are reserved,

whether the whole or part of the material is concerned

specifically those of translation, reprinting, re-use of illustrations,

broadcasting, reproduction by photocopying machine

or similar means, and storage in data banks.

© 1972 by Springer-Verlag Wien
Originally published by Springer-Verlag Wien-New York in 1972

ISBN 978-3-211-81308-9   ISBN 978-3-7091-2440-6 (eBook)
DOI 10.1007/978-3-7091-2440-6

This text has been submitted "ready for camera" by Professor Manfrè and has been reproduced without any corrections or additions, except for the page numbering and the table of contents.

for CISM
Prof. Giuseppe Longo
Responsible for the Editorial Board

LIMIT OF THE SPINNING PROCESS IN MANUFACTURING SYNTHETIC FIBERS

First Part : Phenomenological analysis

Second Part : Experimental and quantitative analysis

Dr. Giovanni Manfré - R & D Division

TECHNION SpA - Novara (Italy)

The present two papers have been delivered in the International Meeting on 'Experimental Methods in Mechanics" held at the International Centre for mechanical sciences - Udine (Italy) on 24-29 October 1974.

Both papers approach the spinning of synthetic fibers in order to help the production engineers to face the process rationalization especially in regards to the limits of rate of flow per nozzle of the spinneret.

Synopsis

The present work deals with the production limits of the spinning process which concerns the manufacture of organic and inorganic fibers included the metals.

The main purpose of the work is to give the technical information necessary to help the production engineers in the rationalization and optimization of the spinning process.

The work is divided in several parts and this paper treats mainly the production limits both in the region of extrusion and drawing zone downwards the bushing.

For the sake of clarity this topic has been divided in two papers.

The present paper intends to be a phenomenological treatment and it will be followed by the next which will deal the same matters from the experimental and calculation point of view.
The production limits are mainly intended a limitation of the rate of flow per one nozzle.

The limits in the bushing region are essentially due to three phenomena:
- die swelling
- capillarity
- melt fracture

The die swelling limits mainly the geometry of the nozzle and the stretching spinning ratio, the capillarity limits the upper temperature of the process and the melt fracture limits the upper rate of flow.

The three phenomena, treated from the rheological point of view, need to be investigated with a basical knowledge of the involved elastic parameters in addition to the more known viscous features of the spun materials.

This implies to find relationship between the normal pressure and the shear rate involved in the liquid flow through a capillary, especially for the polymers.

For this the paper gives a general information of the up to date approaches to the problems and indicates those more suitable for the spinning process.

Introduction

To produce a fiber is simply to transform a given material in a unidimensional shape.

By convention it has been established that 'fiber' means any filament at solid state with length/diameter ratio greater than 10 micron and diameter not larger than 250 micron.

From the geometrical aspect, the fiber can be distinguished in continuous and discontinuous.

The continuous are generally called filaments; the metallic fibers are called microwires.

From the chemical aspect the fiber can be distinguished in organic and inorganic. The former and the latter can be artificial or synthetic.

From the physical aspect, the fiber can be distinguished in one phase or biphase or more phase fiber.

From the structure point of view, the fiber can be distinguished in monocristalline (whiskers), polycristalline, vitreous, non cristalline, and amorphous.

The structure orientation depends on material and process.

The structure and features of fibers depend not only on the intrinsic parameters but also on the process and for the same process on the extrusion and spinning condition and in many cases on the additional treatments.

To produce fiber the processes can be summarized:
- natural (wool, silk, cotton, linen)

- conversion (carbon, graphite, silicate, and allumina)
- cold drawing (metal, alloys and more recently some plastic material like the high density polyethilene)
- the vapour phase deposition and growing (boron, silicon carbide, and whiskers)
- jet spinning from melt or solution through a nozzle with a solification process simultaneous to the fibre stretching. The solidification can be by cooling (nylon polypropilene, polyester, glass), by evaporation of solvent 'dry spinning' (cellulose acetate, polyacrylonitrile, polyvynilcloride) and also by coagulation 'wet spinning' (polyacrylonitrile, polycellulose, polyxantate).

Spinning Process

Nearly all the textile fibers, now available, are industrially produced by the spinning from melt or solution which in other words, means to force a material through a small die to form a free liquid jet at the exit.

This solidifies as it proceeds along the spinning path and the solid fiber is collected on a rotative drum.

Solidification is due to cooling in the melt spinning, to evaporation of a solvent in the dry spinning or to coagulation of polymer from solution in the wet spinning.

The present paper concerns only the rheological behaviour of a forming fiber for the spinning process from melt mainly for polymers but also glass and metals.

This treatment can be extrapolated to concentrate polymer solution by changing duly the temperature parameter with the concentration and the phenomena concerning the heat transformation in phenomena concerning the mass transportation.

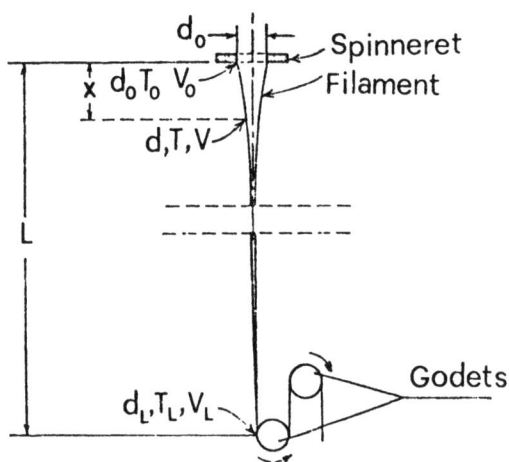

Fig. 1 Scheme of the melt-spinning process.

The spinning process, as shown in fig. 1, can be divided in four zones:
- melting zone
- extruding or bushing zone
- drawing zone or cone zone
- solidifing zone and additional other stretching treatment zones.

In the melting zone, the rheology concerns materials behaviour during the melting and the transport process to the spinneret by equipment like screw extruders, driving drums or gas pressure.

Rheologically speaking, the problems in this zone deal with facts that materials have to get the spinneret completely relaxed.

Furthermore before the spinneret zone, the melt has to be refined, omogeneous and without degradations.

In the extrusion zone, the rheology concerns the phenomena occurring to the liquids at the inlet, inside and at the exit of the nozzle.

The jet just outside the exit of the nozzle is called the drawing zone, a characteristic zone of a spinning process with its typical conical shape, where the main jet attenuation occurs.

Fig. 2a shows the drawing zone of a non-newtonian material.
Fig. 2b shows the drawing zone of a newtonian material.

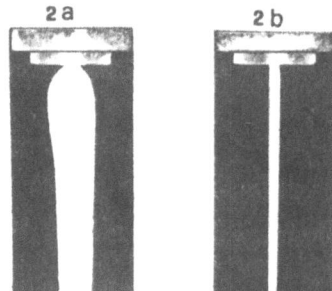

Fig. 2a and 2b show a capillary non-newtonian and newtonian emerging jet from a capillary.

The so called 'solidified zone' can be defined from the point at which some phase transformation occurs up to the tangential point of the forming fiber at the drawing drum.

Although some production limitation occurs also in the solidified zone, generally the main limitations are at the exit of the extrusion spinneret and in the drawing zone.

So the present paper concerns mainly the limitation of the production in these two zones.

Production limits in the extrusion zone

Our productivity limits can be mainly intended in terms of rate of flow limitation per unit nozzle of the spinneret.

We have recognized three main causes of these limits which can be related to:

1 - die swelling effect which consists of a broadening jet emerging from a nozzle. The maximum D_M can be in an extreme condition several time as large as the nozzle diameter D_o (see Fig. 2);

2 - oscillation of the drawing zone in the sense that the geometry of the drawing cone varies periodically in shape and volume;

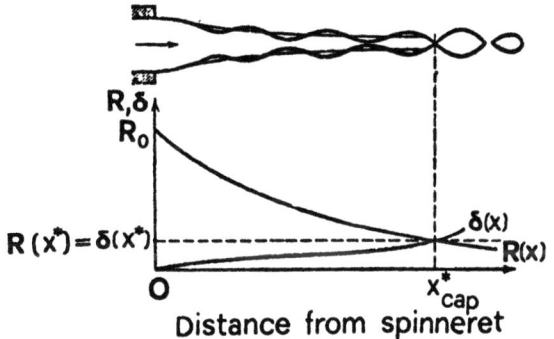

Fig. 3   Capillary breakage of a liquid jet.
The surface wave amplitude $\sigma(x)$ causes the breakage when $\sigma(x)$ equals the corresponding attenuated jet radius $R_o(x)$.

Fig. 3 shows this phenomena which is usually named 'capillary effect'.

This is related to surface wave propagation phenomena along the jet; the breakage of the jet occurs when the wave amplitude $\delta(x)$ becomes equal to the attenuating jet radius $R(x)$.

3 - Melt fracture which occurs when some critical extrusion conditions are exceeded either in terms of pressure or rate of flow (see fig. 4)

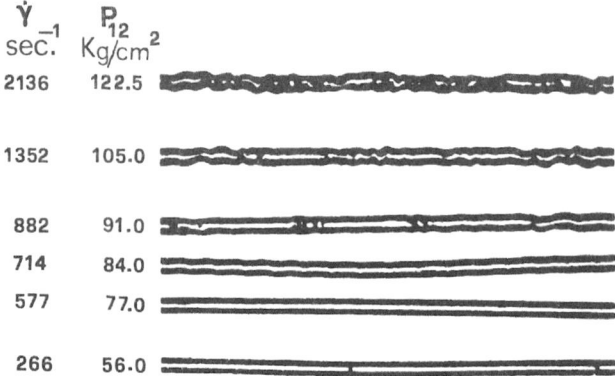

Fig. 4 Distorsion of polymer jet above the critical shear rate (or shear stress) at which melt fracture starts occurring.

This effect leads to rough regular or irregular discontinuity of the extruded material surface.

Usually the melt fracture starts at certain shear stress or shear rate which cannot be exceeded in the spinning extrusion conditions to assure the process stability.

These three phenomena are undesirable in fiber spinning.

In fact <u>die swelling</u> is supposed :
- to affect the uniformity of fiber diameter;
- to give limitation to elongational stretching ratio which changes from D_o/D_f to D_M/D_f being D_f the final diameter of the forming fiber;
- to compel the spinneret designer to take into account the distance between one nozzle and another.

  In fact in case of fiber breakage, there should be interference between the swollen jet and the other adjacent forming fibers;
- to change the final features of fiber.

  In fact it can be shown that the die swelling ratio affects the whole shape of the drawing zone involving differences in temperature and shear rate distribution along the spinning path.

  Because these two distribution are related with the structural orientation, the final fiber features can undergo variations.

<u>Capillarity</u> phenomena certainly is supposed :
- to affect the stability of the process. In fact the oscillations of the drawing zone implies variations of rate of flow (and so of final fiber diameter), of temperature and shear rate distributions which determine the final features of the fiber ;
- to limit the upper temperature (apart the polymer degradation considerations).

  In fact we will show that the capillarity phenomena is related to the viscosity η surface tension α ratio.

Because the viscosity remarkably depends on the temperature, it follows that in the spinning process the capillary phenomena is not important up to a certain temperature range at which some breakage of the drawing zone can occur.

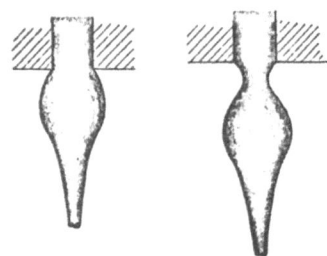

Fig. 5 Breakage at drawing zone due to capillarity effect

The <u>melt fracture</u> is certainly the more evident limitation of the spinning rate of flow.

In fact for the stability of the process and the uniformity of the fibre features, it is necessary to operate in the range of shear stress, or shear rate, quite lower than the critical shear stress at which the melt fracture appears nevertheless with a very low frequency.

The critical melt fracture shear stress has been determined experimentally for any commercial common polymer and it should be given nearly as a parameter of the treated material.

It is worth ending this part by saying that die swelling does exist in any spinning process which, on the other hand, is limited in ranges of temperature and shear stress both lower than the corresponding critical value.

The die swelling and the melt fracture limits are very

much determined by the elastic behaviour of the materials.

To increase the critical limits of the die swelling and the melt fracture it is necessary or to change the geometry of the nozzle or to increase the temperature.

The temperature increasing, degradation apart, falls in the mentioned capillary phenomena.

The relationship among the nozzle geometry, the material intrinsic parameters and the process variables is actually the main topic of the rheologist who needs to rationalize and optimize the spinning process.

This rationalization has to determine the relationship among the intrinsic materials, process and boundary conditions represented in this paper by the symbols listed above.

Symbols *

D_o, R_o = diameter, radius of the nozzle (cm);
D_M, R_M = maximum diameter, radius of the swollen jet (cm);
e = total end effect;
F = tensile force due to winding machine (dynes);
G = elastic shear modules (dynes/cm^2);
l_o = length of spinneret channel (cm);
n = viscosity index of the polymer power law;
n_v = Couette end effect;
P = pressure (dynes/cm^2);
$P_{xx} = P_{11}$ = axial normal stress component (dynes/cm^2)
P_{11}^o = " " shear component at X = 0 (dynes/cm^2);
$P_{xr} = P_{12}$ = shear stress component (dynes/cm^2);
$P_{rr} = P_{22}$ = radial normal stress component (dynes/cm^2);
$P_{\psi\psi} = P_{33}$ = neutral normal stress component (dynes/cm^2);
ΔP = pressure difference along the nozzle (dynes/cm^2);
Q = volume rate of flow (cm^3/sec.);
r = radial coordinate (cm);
R(x) = fluid jet radius function of x (cm);
R_o' = $(dR/dx)_o$ initial derivation at X = 0;
S_r = recoverable shear strain;
t = time and mainly transit time in the nozzle (sec);
t^* = relaxation time (sec);
T = temperature (°C)
V = velocity in the X direction (cm/sec);

* The following symbols are used in both the present papers.

V_o = average extrusion velocity (cm/sec);

x = axial coordinate;

x^* = liquid jet length before breaking for the capillary effect (cm);

a = surface tension (dynes/cm);

$\beta = D_M/D_o$ = ratio due to die swelling;

$\dot{\gamma}$ = shear rate (sec.$^{-1}$);

$\dot{\gamma}_w$ = shear rate at capillary wall (sec.$^{-1}$);

δ = distorsion amplitude of the surface wave (cm);

η = shear viscosity dependent on shear rate (poises);

η_a = apparent viscosity;

ρ = density (g/cm^3);

ξ = $d \log V/dx$ axial deformation gradient;

σ_1, σ_2 = normal stress coefficient dependent on shear rate (dynes/cm^2)$^{-1}$

ϕ = angle coordinate;

$\psi = \overline{V^2}/\overline{(V)}^2$ = function of radial velocity distribution.

Theory

In order to explain the fondamental relationship among the main parameters, some explanation is necessary on the rheological behaviour of the spinning fluid through the nozzle.

The pressure difference ΔP for obtaining the assumed volume rate of flow Q has to be related to the nozzle geometrical dimensions and the intrinsic material parameters.

Assuming that the flow is :
- independent of time;
- laminar, $V_r = V_\phi = 0$;
- fully developed;
- isothermal;
- non slipping at the nozzle wall;
- gravity forces neglecting;

the solution of the hydrodynamic equation of motion for an incompressible, general fluid /1/,/2/, flowing through a cylindrical capillary with radius R_o and length l_o is as follows:

$$V_x(r) = V_x(0) \left[1 - \left(r/R_o \right)^{n+1/n} \right] \quad (1)$$

$$P_{12} = \frac{\Delta P r}{2 l_o} \quad (2)$$

$$P_{11} - P_{22} = -\sigma_1 P_{12}^2 \quad (3)$$

$$P_{11} - P_{33} = -(\sigma_1 - \sigma_2) P_{12}^2 \quad (4)$$

$$Q = \frac{n}{3n+1} \pi R_o^3 \left(\frac{\Delta P R_o}{2 \eta_o l_o} \right)^{1/n} \quad (5)$$

with the hypothesis that the viscosity dependence on rate of shear is the actual power law:

$$\eta(\dot{\gamma}) = \eta_0 (\dot{\gamma})^{n-1} \qquad (6)$$

with η_0 the viscosity of $\dot{\gamma} \rightarrow 0$ (newtonian part of the flow curve η versus $\dot{\gamma}$ at constant temperature T).

Just to fix some idea, it can be shown that for a newtonian fluid n = 1 and so

$$Q = \frac{\pi R_0^4 \Delta P}{8 \eta_0 l} \qquad (7)$$

which is the well known Poiseuille law. Further the velocity distribution changes with n as shown in fig. 6 and it is directly determined by the experimentally obtained flow curves similar to that shown in fig. 7.

It is commonly accepted experimental fact today that for polymeric melts and solutions, σ_1 is positive.

There is, however, less agreement on the sign of σ_2, although it appears that an increasing number of investigations report that it is negative [3],[4],[5], and it has a magnitude of approximatively one or two tenth of σ_1.

It should be noted at this point that both σ_1 and σ_2 can be experimentally calculated with a Weissemberg rheogoniometer:cone-plate geometry for σ_1 and plate-plate geometry for σ_2.

The error can be higher for σ_2.

This consideration apart σ_1 implies an extra tension in the x direction, whereas a negative σ_2 implies an extra tension in the x_3 direction normal to x direction.

At this point, should be noticed that the intrinsic rheological parameters are not only the viscosity, like in the classical hydrodynamics, but the viscosity η_0, the index n, the elastic constants σ_1 and σ_2.

So the tensor for the steady state flow can be as follows:

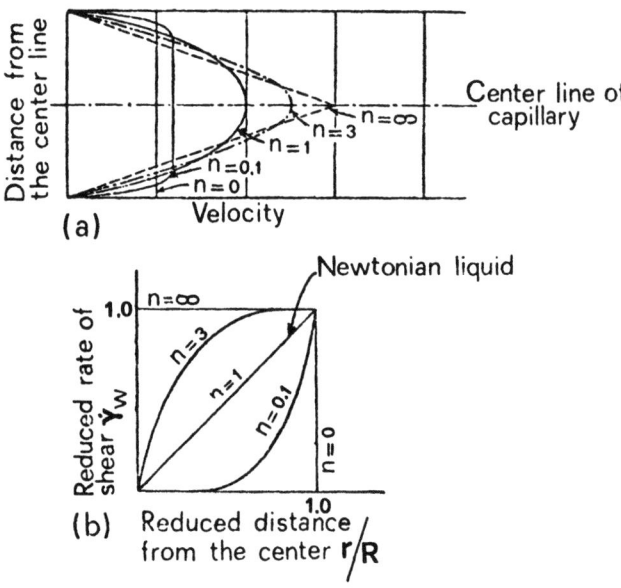

Fig. 6 Velocity distribution and reduced rate of shear in function of viscosity index for a capillary flow.

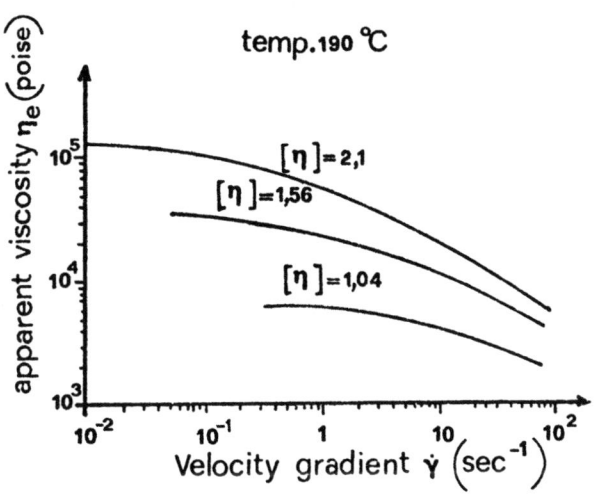

Fig. 7   Apparent viscosity versus velocity gradient for polypropilene at temperature 190 °C and 3 different intrinsic viscosity.

$$P = -P_0 I + \begin{Bmatrix} \sigma_1 \eta^2 \dot{\gamma}^2 & \eta \dot{\gamma} & 0 \\ \eta \dot{\gamma} & \sigma_2 \eta^2 \dot{\gamma}^2 & 0 \\ 0 & 0 & 0 \end{Bmatrix} \quad (8)$$

where I is the unitary tensor, P_0 the ambient pressure, η the non newtonian apparent viscosity.

When $\dot{\gamma}$ increases η normally decreases 1/10 - 1/30 from its original value η_0 and n varies from 1 to 0,1.
For this the formula (5) indicates that the rate of flow from the nozzle is actually larger than that calculated with the Poiseuille law (7).

As $P_{11}-P_{22}$, $P_{11}-P_{33}$ are surely related to the elastic phenomena it is worth to point out that not only the value of η is in function of molecular weight, its distribution and other more sophisticated structural parameter /7/ but also σ_1 and σ_2.

Concerning σ_2 we have only very scanty data and for the sake of simplicity we can say by now that the possible influence of $P_{11}-P_{33}$ in our following treatment can be nearly the same that $P_{11}-P_{22}$.

This implies the validity of the mentioned experimental results showing that σ_2 tends to be at least one tenth of σ_1 and so that $P_{22}-P_{33}=0$.

Consequently, at least for rate of shear lower than that corresponding to melt fracture, what we shall show for $P_{11}-P_{22}$ can be also extrapolated to $P_{11}-P_{33}$.

It has been shown that in case of stationary motion, this practically means V_R more than 20 (see fig. 8).

For concentrated solution of polymers /8/ we have:

$$\sigma_1 = 2 J_0 \qquad (9)$$

$$J_0 = \frac{2.2 \, J_R}{1+ 2.1 \times 10^{-5} C M_w} \qquad (10)$$

$$J_R = \frac{2}{5}\left(\frac{M_w}{CRT}\right)\left(\frac{M_z M_{z+1}}{M_z^2}\right) \qquad (11)$$

J_R is a steady state shear compliance according to Rouse theory /9/; M_w, M_z; M_{z+1} are average molecular weights defined as ratios of the moments of the molecular weight distribution; C is the polymer concentration (gm/ml); R is the gas constant and T is the absolute temperature.

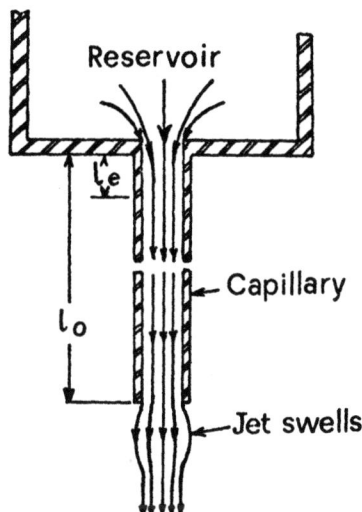

Fig. 8 Stream lines of stationary liquid flow with the capillary with length l_o much longer than the critical. Up to the length l_e the flow is affected by the entrance effect.

Note that σ_1 decreases as temperature T or concentration C increases while σ_1 increases with the molecular weight M_w and with the broadening of molecular weight distribution.

Formula (9) has been experimentally verified above all for low shear rates, also for polymeric melts.

From this, it can be established a relationship between the $P_{11} - P_{22}$ and the die swelling ratio β, either knowing $P_{11} - P_{22}$ by rheogoniometer measurement for calculating β or knowing experimentally β for calculating $P_{11} - P_{22}$ /10/.

This part can be concluded, noting that the die swelling effect with capillary ratio $l_o/R_o > 30$ and low shear rate can be caused by P_{11} and P_{22} originated by the steady shear flow inside the capillary.

In other words the elastic deformation, originating during the capillary flow can undergo some relaxation because the transit time of the flowing polymer $t = \pi R_0^2 l_0 / Q$ is higher than the Maxwell relaxation time $t^* = \eta/G$ of the polymer itself which usually is of the order of 0,1 second in the normal polymer spinning conditions.

When the stationary state conditions are not satisfied, that is when shear rates are relatively high or l_0/R_0 are low, the die swelling phenomena are not explained by the model shown in (9).

This is mainly due to the elastic energy, stored at the entrance capillary, which has not enough time to relaxe along the capillary (ratio t/t^* nearly 1 or less than 1).

So the polymer melt approaches the nozzle exit like an elastic solid ready to recover its initial unsheared state by swelling at the die exit.

This model which considers the polymer a Hooke-like solid can be described by the theory of rubber elasticity which involves the the shear elastic modulus G which can be calculated simply by the formula :

$$G = \rho RT / Me \quad (12)$$

with Me the average molecular weight of the polymer.

Thus it can be shown that in this model

$$\sigma_1 = \frac{1}{KG} \quad (13)$$

with K predicted by certain theories for elastic liquids and experimentally found $K = 1/2$ by Lodge [11] or $K = 1$ by Weissemberg [12].

A value of K between this value is usually observed experimentally.

Because the l_0/R_0 ratio in the industrial spinning process ranges between 2 and 8, it appears more satisfactory to relate $P_{11}-P_{22}$ with die swelling data by means of the second model either knowing $P_{11}-P_{22}$ for calculating β or knowing β experimentally for calculating $P_{11}-P_{22}$. Notwithstanding equation (13) cannot be regarded as completely verified (14) it is nevertheless employed in the spinning process rationalization to obtain an approximate relation between the recoverable shear strain and the stresses involved in shear capillary flow.

Conclusion

This part has analyzed, from the phenomenological point of view, the main sources of instability which limit the spinning process in the extrusion zone.

The die swelling, the capillary and the melt fracture phenomena can be indicated the three phenomena which limit the geometrical size of the nozzle, the upper temperature and the rate of flow of the spinning process.

The second part of this work will concern the quantitative calculation which should lead to predict and rationalize the mentioned limits.

In order to approach properly the second part a theoretical treatment has been described in order to point out the intrinsic parameters, the process variables and their possible relationship to be used.

Particular emphasis has been devoted to the normal stresses involved and their relationship with the intrinsic elastic constants which can be related also to temperature and molecular weight with its distribution.

Among the possible models necessary to establish the more actual relationship for the spinning process, the recoverable shear elastic strain at the nozzle exit, seems to be the suitable main parameter.

The second part will deal with extrusion and spinning experiments necessary to calculate properly this parameter which leads the engineers to rationalize the process in the extrusion zone and to design the process in case they are spinning a new material, included glasses and molten metals.

References of first part

/1/ LODGE A.S.,'Elastic Liquids' Ed. Academic Press -London (1964) 214-220

/2/ MARK H.F.,ATLAS S.M.,CERNIA E.,'Man made fibers' vol. 1 Ed. ZIABICKI 36 - 47 Ed. Interscience Publishers - New York (1967)

/3/ PRITEHARD W.G., Phil. Trans. Royal Soc. $\underline{270}$; 507 (1971)

/4/ HIGASHITAMI K.,Trans. Soc. of Rheology, $\underline{16}$-687 (1972)

/5/ GINN R.F., METZER A.B., Trans. Soc. of Rheology $\underline{13}$,429 (1969)

/6/ TANNER R.I., Trans. Soc. of Rheology $\underline{17}$, 365 (1973)

/7/ BUECHE F.,HARDING S.W.,J. Poly Science $\underline{32}$, 177 (1958)

/8/ GRAESSLEY W.W., Trans. Soc. Rheology $\underline{14}$, 519 (1970)

/9/ FERRY J.D.,'Viscoelastic properties of polymers' Ed. Wiley Interscience, New York (1970)

/10/ SPEAROT J.A.,METZER A.B.,Trans. Soc. of Rheology $\underline{16}$ - (3) 495 (1972)

/11/ LODGE A.S., Rheol. Acta $\underline{1}$, 158 (1958)

/12/ WEISSEMBERG K., Nature $\underline{159}$, 310 (1947)

/13/ POLLET W.F.O., Proceedings Second Int. Rheological Congress, Butterworth, London (1945)

/14/ KELLEY F.,BUECHE F., J. Poly Science $\underline{50}$, 549 (1961)

LIMIT OF THE SPINNING PROCESS IN MANUFACTURING SYNTHETIC FIBERS

Second Part : Experimental and quantitative analysis

Dr. Giovanni Manfré - R & D Division
TECHNION SpA - Novara (Italy)

Synopsis

The main purpose of this paper is to describe the experimental work necessary to help the production engineers to make quantitative considerations on the synthetic fiber production limits of the spinning process.

The paper follows the phenomenological treatment presented earlier [1].

The die swelling, capillarity and melt fracture are considered the main sources of the production limits in terms of rate of flow per one nozzle of the spinneret.

The three phenomena depend on geometrical shape of the nozzle, process parameters, rheological and surface material properties. Thus they can be related to the so called 'end effects' occurring in any viscous flow through a capillary.

The end effects are divided in kinetic and elastic effects and can be experimentally found.

Their interpretation leads to the relationships between stresses and shear rate necessary to know in order to predict the production limit and so to rationalize the spinning process in the extrusion and drawing zones.

Some comparison between polymers, glasses and metals from melt spinning point of view facilitates the knowledge of the phenomena, showing the utility of the presented calculation and finally it can help facing problems to be solved in the spinning new materials from melt or solution.

In fact this approach led our work to approach the spinning of metals from melt.

This treatment has to be completed by further work concerning the limitation of production involved along the path of the forming fibers, where phase transformation can also occur.

Introduction

The first part of this work [1] dealt with the phenomenological view related to the production limits of the synthetic fiber spinning process.

The main effort has been devoted to indicate that the die swelling, the capillary effect and the melt fracture can be considered the main sources of the production limits.

The die swelling and the melt fracture are surely related to the elastic properties of the spinning materials which, in the extrusion of the spinning process, play an important role, mainly due to the involved high shear stresses and small capillary ratios l_0/R_0.

For this, the normal stresses P_{11}, P_{22}, P_{33}, which are directly related with the material elastic constants, have been particularly treated in order to prepare the reader how they can be used to investigate and calculate the production limits through experiments carried out with the actual spinning nozzles.

Among the models, which can lead to relate the die swelling and the melt fracture to the so called 'end effects' we have decided that both depend on the liquid elastic deformation occurred at the inlet of the capillary (elastic shear strain) and inside the capillary due to shear stress (relaxational phenomena).

Both elastic shear strain and relaxational phenomena have

been related [1] with the polymer intrinsic parameters as viscosity η, molecular weight and its distribution and other structural features as the cross-link numbers and the entanglements density [2],[3] and the Maxwell relaxation time $t^* = \eta/G$.

Unfortunately the recoverable shear strain and the relaxation effect occur simultaneously and both can be quite large in the spinning process. In fact the first, mainly related to the l_0/R_0 ratio ranging 2-8 in the spinning process, has no time to relax inside the capillary. Further the ratio t/t^* between the transit time and the relaxation time t^* is very small.

The transit time $t = \pi R_0^2 l_0/Q$ is the average residence time of the fluid within the capillary and in fiber spinning it is of the order of 0.1-100 m/sec. while relaxation times t^* are expected to be of the order of 100-500 m/sec.

So far the experimental approaches to divide the two simultaneous effects also with large l_0/R_0 ratio seem to leave many doubts [4]. But for the sake of simplicity we have decided that the elastic total end effect is mainly due to the recoverable shear strain directly related to the shear elastic modulus G of the spun material as described in formula (13) in [1]

For the interpretation of capillary effect, which is not an actual end effect, a surface tension and surface phenomena have to be involved.

For this, the fluid motion inside the capillary and the liquid flow in the drawing zone are to be bridged.

This means to solve the motion equation with elastic and surface effect on boundary conditions.

The presence of additional external forces, as the stretching force attenuating the fiber forming diameter by the winding machine, adds more complexity to the problem and the resulting given formula have to be taken only for approximate solution.

The validity of the solution for the capillarity effect can have some confirmation in the glass and metal spinning provided to neglect variation temperature effects; thus the solution has to be considered at an average temperature.

Experimental measurements

The experimental work necessary to carry out on the spun material can be splitted in two parts:

a) measurements of the normal stress differences $P_{11} - P_{22}$ and $P_{22} - P_{33}$ versus P_{12}, in order to obtain the value of the elastic constants σ_1 and σ_2. This can be done mainly by experimental investigations in liquid steady shear flow :

- in concentric cylinder (optical phase difference, extinction angle or torque)
- in cone-plate or plate-plate (Weissemberg rheogoniometer);
- by other direct measurements of shear modulus G or average relaxation time $t^* = \eta/G$;

b) rheological measurements with a capillary extruder equipped with temperature control and a series of nozzles exhibiting different l_0/R_0 ratio. The geometrical dimension and shape of the nozzles must be as much as similar to those used in the industrial spinning process.

At any given temperature it is possible to obtain the flow curve, pressure P versus rate of flow Q for each capillary with different l_0/R_0 ratio.

From the experimental curve P – Q it is easy to obtain the plot of apparent viscosity η_e versus the shear rate $\dot{\gamma}_w$ calculated at the capillary wall as it is reported in fig. 7 of /1/ for the polypropilene with different intrinsic viscosity.

The apparent viscosity η_e is calculated as the ratio between the shear stress P_{12} and the shear rate $\dot{\gamma}_w$ at the capillary wall given by:

$$P = \frac{\Delta P R_0}{2 l_0} \qquad (1)$$

$$\dot{\gamma} = 4Q/\pi R_0^3 \qquad (2)$$

which means to have treated the spun material a newtonian-like liquid following the Poiseuille law.

From the log η_a – log $\dot{\gamma}_w$ curve, as shown in fig. 7 of /1/ the viscosity coefficient n can be calculated:

$$\frac{d \log \eta_e/\eta_0}{d \log \dot{\gamma}} = n - 1 \qquad (3)$$

thus the power law can be derived

$$\eta = \eta_0 \left(\dot{\gamma}_w\right)^{n-1} \quad (4)$$

With this data, the velocity distribution in the capillary becomes (see formula (1) in /1/:

$$V_x(r) = V_x(0)\left[1-\left(r/R_0\right)^{n+1/n}\right] \quad (5)$$

so the true shear rate $\dot{\gamma}$ is recalculated by formula (2) with Q now given by:

$$Q = \frac{n}{3n+1} \pi R_0^3 \left(\frac{\Delta P R_0}{2\eta_0 l}\right)^{1/n} \quad (6)$$

Now knowing the true shear rate $\dot{\gamma}$, calculated for capillaries with different l_0/R_0, it is possible to plot the experimental pressure P versus l_0/R_0 for constant shear rate $\dot{\gamma}$ as it is shown in fig. 1 for the polyethylene.

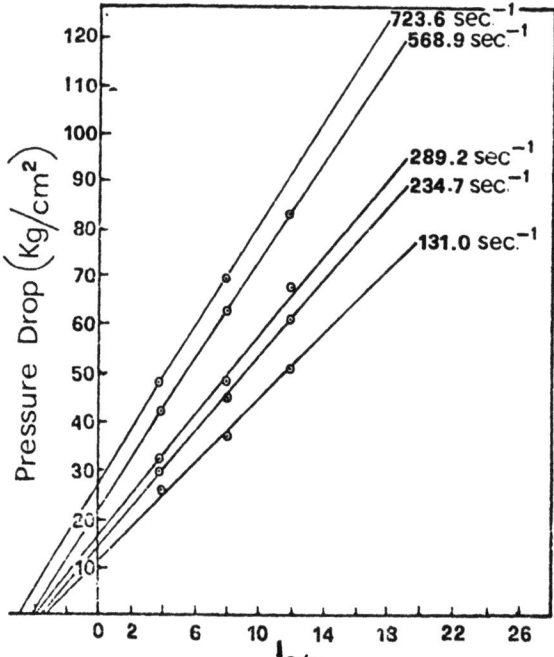

Fig. 1 Bagley plot versus l_0/D_0 ratio for polyethylene at 180 °C.

Generally speaking [6], [7], the experimental curves are similar to that shown in fig. 1, although a more careful investigation presents a nonlinear plot for both l_0/R_0 low and shear rate $\dot{\gamma}$ approaching zero.

This can be due to the interference of different end effects the elastic and the kinematic, in these conditions. But normally, the extrapolation of the experimentally obtained curve lets to calculate the pressure loss P_0 at $l_0/R_0 = 0$ and the total effect e at $P = 0$ for each gradient $\dot{\gamma}$.

The pressure loss P_0 means that along the capillary, the effective pressure drop, associated with shear flow, is not equal to the applied pressure but there exist additional pressure losses due to kinematic effects and elastic effects. These effects can be considered either in terms of additional pressure P_0 or in terms of additional length eR_0 for the capillary. The experimental results, so elaborated, lead to plot the end effect l, or the additional pressure P_0, versus the shear rate $\dot{\gamma}$. Practically for a short range of $\dot{\gamma}$ an average value of n can be found and a true shear stress at the capillary wall can be calculated

$$P_{12} = \Delta P/2\left(\frac{l_0}{R} + e\right) \quad (7)$$

with the pressure drop along the capillary $\Delta P/l_0 + eR$

The end effects

Generally speaking the total 'end effect' e can be splitted

in three parts:

$$e = n_V + S_r/2 + f(t/t^*) \quad (8)$$

where n_V is the kinematic effect or Couette effect, $S_r/2$ the recoverable shear strain due to the entrance elastic effect and $f(t/t^*)$ the 'memory effect' which is a function of transit/relaxation time ratio. It can be shown /8/ that:

$$n_V \simeq \frac{\rho V_m^2 l_o}{R_o N \Delta P} \Big/ 1 - \frac{\rho V_m^2}{N \Delta P} \quad (9)$$

with n_V = average velocity inside the capillary, ΔP the pressure applied drop and N the non-newtonian parameter :

$$N = (4n+2)(5n+3)/3(3n+1)^2$$

Due to the low value of V_m in the usual spinning process, the value of n_V can be considered less than 1% of the total end effect e and so can be neglected for polymers in spinning. This inertial term can assume importance in the molten metal spinning due high velocity and in the glass spinning due to the absence of both elastic terms.

For the separation and the calculation of $f(t/t^*)$ from $S_r/2$ both elastic effects, we have already discussed in /1/.

Our conclusion was that in fiber spinning, due to relatively high shear rate, usually 10÷200 sec^{-1} and low l_o/R_o ratio, the main term is the recoverable shear strain $S_r/2$.

With this hypothesis formula (8) becomes:

$$e \simeq S_r/2 \quad (10)$$

and fig. 2 shows /9/ the recoverable shear strain in function of shear rate $\dot{\gamma}$.

This means that in spinning the kinematic term and the relaxation term can be neglected and the end effect is mainly due to the elastic energy originated and stored by the molten polymer at the inlet of the capillary. This conclusion might be too drastic and can leave some contradiction because, in other words, we treat the spinning liquid as a Hooke-like solid which undergoes elastic strain at the capillary inlet, does not relax along the capillary and it relaxes at the exit, producing the die swelling. Up to a certain critical shear stress S_r varies linearly and starts being constant above a certain stress.

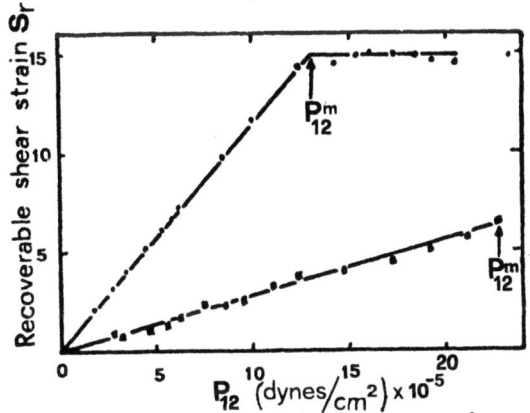

Fig. 2   Recoverable shear strain versus shear stress. P_{12}^m corresponds to critical shear stress for melt fracture.

The stress at which S_r does not increase anymore, coincides experimentally with the range of melt fracture as shown in Fig.2. The linearity of S_r versus the shear stress P_{12} at the capillary wall and the hypothesis described in formula (13) of paper /1/, lead to conclude that :

$$P_{11} - P_{22} = \sigma_1 P_{12}^2 = S_r P_{12} = \frac{G}{K} S_r^2 \qquad (11)$$

Thus can be experimentally related to the die swelling and the same results are described in /10/,/11/.

One of the most useful relationship is

$$P_{11}-P_{12} = \frac{G}{K}S_r^2 = G\left(\beta^2 - \frac{1}{\beta}\right) \quad (12)$$

which can be theorically derived /10/ by simple calculations concerning a cylinder elastically stretched, which enlarges its cross section when the stored elastic strain recovers or, in other words, when a tension $P_{11}-P_{22}$ acts perpendicularly to its axis.

Formula (12), or other similar, has been experimentally verified and they have confirmed with some particular exception like PVC, that the die swelling depends on temperature T, molecular weight and its distribution through G, and on l_0/R_0 through P_{12}.

More particularly, β increases with shear rate, molecular weight and the wideness of its distribution, decreases when l_0/R_0 and temperature increases.

Formula (12) can be mainly applied for polypropilene, polyethylene, polymethylmetacrylate but it can be given only approximate values for nylon, polyester and concentrated polymer solutions /13/.

When some exception occurs, the stated hypothesis must be verified : value of n_V, influence of relaxation term $f(t/t^*)$ and also to have considered negligible the difference $P_{11}-P_{33}$.

Furthermore, concerning the die swelling of polymers, or that of glasses or metals, we have to take into account the external forces as the tensile force produced by the winding machine for attenuating the forming fiber, and the surface tension a.

Die swelling in spinning

Due to the tensile force F and surface tension a the die swelling of the forming fiber (see fig. 3) is smaller than that measured as a simple end effect.

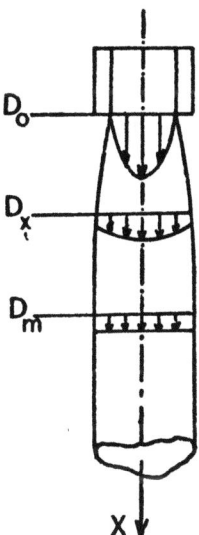

Fig. 3   Die swell with external forces involved in drawing zone. D_m is the relaxed zone.

In fact β can be 1 to 5 for free jet and in spinning can be 1,5 maximum. For this it is possible to make the momentum balance of the exit portion of a free jet, between the exit cross section D_o and the relaxed downstream cross section D_m, where the velocity profile can be assumed uniform and both surface friction in the surrounding ambient and gravity force can be neglected /14/.

The momentum balance, as a result of solution of hydrodinamic equation of motion /16/, leads to the following equation for die swell ratio β :

$$a/R_0 \left[\beta - \left(1+R_0'^2\right)^{-1/2}\right] + \frac{\rho Q^2}{\pi R_0^4}\left(\psi_0 - \frac{1}{\beta^2}\right) - P_{11}^0 = 0$$

For 'power low fluids' ψ_0 ratio between velocity distribution inside the capillary and the average along the jet $V_m = Q/\pi R_m^2$ can be calculated from the viscosity index n as:

$$\psi = 3n + 1/2n + 1$$

this means that β cannot be 1 even for the newtonian fluids.

$P_{11}^0 = P_{11} + P_0$ is taken as the average extra tensile stress in the cross section D_0 and P_0 is the ambient pressure. Experimentally the value of P_{11} is very difficult to calculate but can be expressed in terms of recoverable shear strain as in formula (12). In fact with surface tension effect nearly neglecting ($a \rightarrow 0$) equation (14) becomes formula (12).

According to (14), β increases with P_{11}^0 and decreases with ψ_0 for very small rates $\beta \rightarrow 1$. The effect of surface tension a is always directed to diminish the curvature of free surface. For small β $\left(\beta < \left[1+R_0'^2\right]^{-1/2}\right)$ an increase of β leads to less contraction; for higher β an increase of surface tension a is associated with smaller expansion. For newtonian liquids, like water, $P_{11} = 0$ and $\psi = 4/3$, β lies between $(3/4)^{1/2}$ for $aR_0^3/\rho Q^2 \rightarrow 0$ (contraction due to inertial forces) and 1 for $aR_0^3/\rho Q^2 \rightarrow \infty$ (determining effect of surface tension).

Some considerations can be done with polymers, glass and metals considering that their surface tension is respectively about 20, 300÷400, 800÷1200 dynes/cm.

We can conclude this part saying that in the case of polymers, with a quite small, the elastic effect is the most important.

To reduce then β, it is necessary to change properly:
- intrinsic parameters as molecular weight, its distribution, the number of entanglements;
- geometrical sizes of capillary as l_0/R_0 or its shape, mainly the angle of the inlet;
- temperature up to a value above which the capillary effect can limit the production.

Generally speaking, the temperature effect does not affect consistently the die swelling ratio. In fact the temperature affects exponentially the viscosity but linearly as the surface tension (3-4 dynes for 100 °C) as the shear modulus G which both directly affect β.

Melt fracture

The most evident production limit is the melt fracture, which seems to occur at certain shear stress which, at given temperature, can be considered the upper production limit /17/. Table 1 shows this limit for some known polymer:

Table 1

Polymer	Shear stress (dynes/cm^2)	Temperature °C
Polyethylene	1,5 -2,2 10^6	150-220
Polypropilene	0,8 -1,2 10^6	200-300
Polytetrafluoroethylene	1 - 2 10^6	360
Polymethylmetacrylate	5 10^6	170
Nylon 66	9 10^6	275
Polystirene	1,2 10^6	20

The hypothesis that the melt fracture is determined by elastic effects at certain shear stress (see table 1), above which the recoverable shear strain remains constant, (see fig. 2) leads to the conclusion that it is independent of the l_0/R_0.

This has been experimentally confirmed both for a range of l_0/R_0 ratio between 2-10, which is the actual range used in spinning synthetic fibers, and for cylindrical shape in the nozzle.

In fact the independence of melt fracture from short l_0/R_0 comes from the hypothesis of the elasticity stored at the inlet, which remains unrelaxed at the nozzle exit as the ratio t/t^* is less than 1.

Reducing R_0 the transit time increases but elastic strain at the inlet increases as well; increasing R_0 the transit time t decreases as well.

The only left possibility was to change the shape of the capillary.

The results with taped capillary have achieved higher critical shear stresses for the melt fracture [18],[19],[20].

These investigations have also pointed out that the distorsion of the polymer jet in addition to the elastic effects, depends on phenomena as the slipping and the lack of adhesion between metal and polymer at the capillary wall.

In fact experiments with tapered die with half angle of $2°-10°$ led to increase 10 times the shear rate.

The metal - polymer interface adhesion as a source of distorsion, see fig. 4, has been found experimentally also using different metal for nozzles.

Its interpretation can be due to the fact that the elastic recovery, introduced during the acceleration at the die entry, is triggered by slipping in the die at the polymer metal interface in the direction opposite to that of the flow.

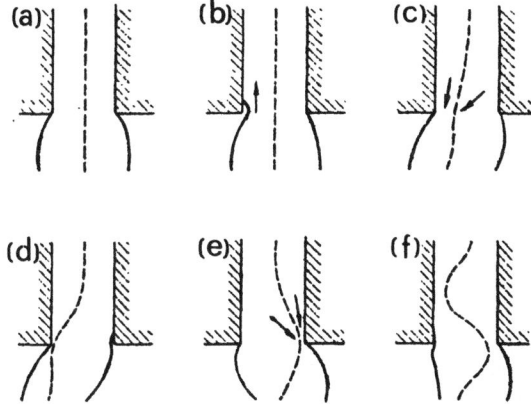

Fig. 4 - Distorsion mechanism at melt fracture
    a - Streamline flow;
    b - recovery from die exit due to slipping;
    c - preferential flow;
    d - recovery from die exit at opposite side;
    f - site progressing alternatively.

The initial site for this triggering may be, at or near to the die exit, where adhesion is lowest.

The effect occurs alternatively on one site and another as shown in fig. 4, and so it produces the periodicity of the distorsion.

The possibilities of improving die design, have in no way been exhausted and it may be possible to exceed the already obtained advantages in production conditions. Unfortunately, the most suitable die for each material must be determined experimentally.

Capillarity instability in spinning

As the die swelling and the melt fracture limit the geometry of the capillary and the spinneret as the capillary limit the upper temperature and the final features of the fiber.

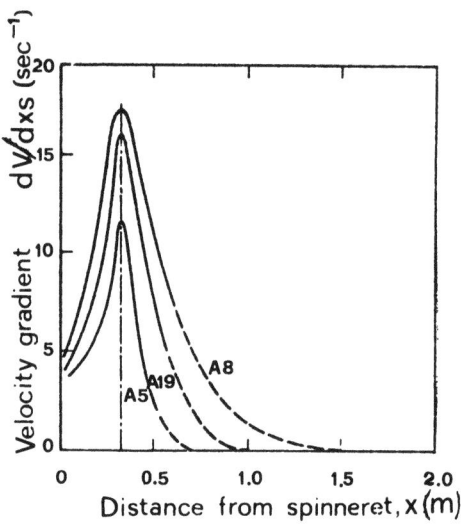

Fig. 5 - Elongational shear rate dV/dx along the forming fiber in the drawing zone A_5, A_{19}, A_8 different final speed.

The capillary instability can be due to any source; pressure and rate of flow variations which produce transversal or longitudinal oscillations of the drawing zone. A too high temperature reduces too much the viscosity and then at surface jet,wave oscillation can propagate from the die exit. The drawing zone becomes unstable and a breakage of fiber can occur before the solidification process takes place. This can be avoided through increasing the elongational viscosity.

It must also be noticed that, as shown in fig. 5, in this particular region the maximum shear rate of the attenuating fiber can occur.

Thus the less viscosity at the maximum shear rate, simultaneous to wave surface propagation, surely facilitates the breakage of the fiber.

Furthermore any oscillation of the drawing zone lead to a variation of its temperature distribution /21/ and thus a variation of the final molecular orientation which can be described by fig. 6

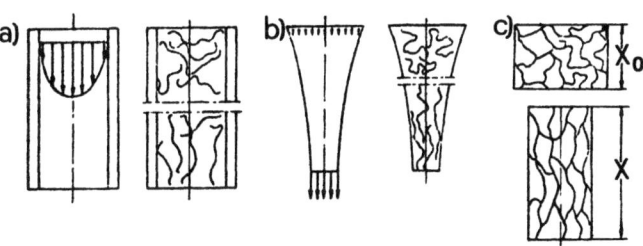

Fig. 6 - Scheme of orientation mechanism;
    a - streaming orientation in shear flow inside the capillary;
    b - streaming orientation in drawing;
    c - deformational orientation in solid zone.

Variation of molecular orientation certainly produce variation of fiber features.

Referring to fig. 3 of paper /1/ and without entry mathematical details, the capillary effect can be related to the so called 'spinnability' property of material.

Its measurements consist to know the maximum liquid jet length x^* at which the first breakage occurs.

This length can be calculated as:

$$x^* = 2A^{-1}\left[\ln R_0/\sigma_0 - a/3\eta \, AV_0R_0\right] \quad (15)$$

with $A = d\ln V_x/dx$ is the deformation gradient along the fiber attenuation, $\sigma_0 =$ initial amplitude of surface wave which can be related to die swell ratio β taking simply $\sigma_0 \simeq \beta$ during the spinning process and V_0 the average velocity at the nozzle exit.

Assume that along the attenuation the velocity changes with an exponential law $V = V_0 e^{Kx}$ and that the fiber breaks at x^* where $\sigma(x^*) = R(x^*)$.

Thus knowing the spinning parameters K, β, η, a and $V_0 = Q/\pi R^2$ one can achieve an approximate value of x^*.

In fact with the above assumption formula /21/ becomes:

$$x^* = 2/K\left[\ln \beta - \pi R_0 \, a/3\eta KQ\right] \quad (16)$$

K depends on many processes and rheological variables but the main are temperature, viscosity-temperature coefficient, stretching force F, and elongational viscosity λ (only in the case of newtonian liquid can be taken 3η as we assumed in formulas /21/ and /22/.

It can be seen that, taking constant the process conditions, the stability of the spinning (longer x^* more stable the process is) depends mainly on the a/η ratio.

In spinning conditions we can assume that for the polymers we can consider this ratio as $20/10^3 = 2.10^{-2}$, for the glasses $300/10^3 = 3.10^{-1}$, for the molten metals $800/10^{-1} \div 1200/10^{-1} = 8.10^3 \div 1,2.10^4$.

By this consideration we can conclude that polymers (drawing zone about 10-50 cm) are 1 order of magnitude more spinnable than glasses (drawing zone about 2 cm) and 5 order of magnitude in regards to molten metals.

Through this calculation we have approached the glass spinning investigation /22/ and also the molten metal spinning /23/.

The results obtained with a new molten metal spinning process have more and more convinced that the theory and the calculation presented in this paper can be used with success to rationalize any spinning process.

Conclusion

After this first part, concerning the phenomenological view, this second part deals with an approach to rationalize the three phenomena :
- die swelling, melt fracture, capillarity.

These can be taken the main sources of instability and so the limitation of the spinning process at least concerning the extrusion and the drawing zone.

The die swell can be calculated through the experimental value of the recoverable shear strain which is also in function of elastic modulus and so polymer molecular weight, its distribution and temperature. The die swell has been treated also taking into accqunt also the external forces: attenuating force from the winding machine and the surface tension.

The melt fracture has been treated considering also the effect of the shape capillary which seems to affect the critical shear stress values. More experiments need to optimize both the best angle of tapered nozzle and materials for the spinneret.

The capillary effect has been related with the so called 'spinnability properties' of the liquid.

The oscillation of the drawing zone can be predicted in function of the ratio between surface tension and viscosity of liquid.

This investigation has led our research to approach the rationalization of the glass spinning and also the possibility of spinning molten metals.

The results, obtained so far, strengthen more and more our opinion that this approach could help any engineer involved in spinning process.

Further work intends to deal with other sources of spinning limits along the solid zone in order to complete our investigation in this matter.

References

/1/ MANFRE' G.,'Limits of the spinning process in manufacturing synthetic fibers'1st part - Phenomenological analysis Further paper here presented.

/2/ GRAESSLEY W.W.,SEGAL L., Macromol., $\underline{2}$, 49, (1969)

/3/ GRAESSLEY W.W.,PRENTICE J. , J. Polymer Science, $\underline{6}$, 1887 (1968)

/4/ KAST W., Kolloid 2, $\underline{187}$, 89,(1963)

/5/ KAYE A.,LODGE A.S. ,VALE D.G. , Rheol. Acta $\underline{7}$, (4) 368,(1968)

/6/ BAGLEY E.B. ,J. Applied Physics, $\underline{28}$, 624,(1957)

/7/ HAN C.P.,CHARLES M. , Trans. Soc Rheol. $\underline{15}$,(2) 371,(1971)

/8/ VAN WAZER J.R., LYONS J.W., KIM K.Y. COLWELL R.E.'Viscosity and flow measurements' 200-210 Ed. Interscience Publishers New York (1963)

/9/ BAGLEY E.B., Trans. Soc. Rheol. $\underline{5}$, 355,(1961)

/10/SPEAROT J.A., METZER A.B., Trans. Soc. Rheol. $\underline{16}$, 495,(1972)

/11/VLACHOPOULOS J.,ET OTHERS ,Trans. Soc. Rheol. $\underline{17}$,669,(1972)

/12/CAPPUCCIO V.,COEN A.,BERTINOTTI F.,CONTI W., Chimica e Industria $\underline{44}$, 463, (1962)

/13/PAUL D.R., J. Applied Polymer Science $\underline{12}$, 2273,(1968)

/14/MANFRE' G. ,Glass Technology $\underline{10}$, (4),99,(1969)

/15/ZIABICKI A., KADZIERSKA K.,Kolloidzeitschrift, $\underline{171}$, 51, (1960)

/16/FREDRICKSON A.G. ,'Principles and Applications of Rheology Ed. Prentice Hall Englewood Cliffs New Jersey (1964)

/17/TORDELLA J.P. , J.Applied Physics $\underline{27}$, (5), 454,(1956)

/18/COGSWELL F.N. ,LAMB P. , Trans.J.Plastics Instr. 809, (Dec.1967

[19] BENBOW J.J., LAMB P., SPE Transactions, 7, (Jan. 1973)

[20] SCHULKEN R.M., SPE Journal 423, (April 1960)

[21] MANFRE' G. Rheol. Acta <u>12</u>, 349, (1973)

[22] MANFRE' G. Verres refractaire <u>26</u>, (2), 57, (1972)

[23] MANFRE' G., SERVI G., RUFFINO C., J.Mat. Science, <u>9</u>, 74, (1974)

CONTENTS

	Page
First Part: Phenomenological analysis	3
Synopsis	4
Introduction	6
Spinning Process	8
Production limits in the extrusion zone	11
Symbols	16
Theory	18
Conclusion	26
References of first part	28
Second Part: Experimental and quantitative analysis	29
Synopsis	30
Introduction	32
Experimental measurements	34
The end effects	37
Die swelling in spinning	41
Melt fracture	43
Capillarity instability in spinning	46
Conclusion	50
References	52

MIX
Papier aus verantwortungsvollen Quellen
Paper from responsible sources
FSC® C105338

If you have any concerns about our products,
you can contact us on
ProductSafety@springernature.com

In case Publisher is established outside the EU,
the EU authorized representative is:
**Springer Nature Customer Service Center GmbH
Europaplatz 3, 69115 Heidelberg, Germany**

Printed by Libri Plureos GmbH
in Hamburg, Germany